3D Printing Introduction Guide

Getting Started with 3D Printing to Help you make Passive income for your Business

by *Jeff Heldrich*

Copyright @ Jeff Heldrich 2018
Cover Design @ Pgrafis

The author and publisher have made every effort to ensure that the information in this book was correct at press time. Although the author and publisher do not assume, and at this moment, disclaim any liability to any party for any loss, damage, or disruption caused by errors or omissions, whether such errors or omissions result from negligence, accident, or any other reason. Before practicing the skills described in this book, be sure that your equipment is well maintained, and do not take risks beyond your level of experience, aptitude, nor training.

(978) 750-8400,
fax (978) 646-8600,
or on the Web at www.copyright.com

All rights reserved.

Table of Contents

Introduction: Benefits of 3D manufacturing 4

How does 3D printing work? 7

Is 3D printing profitable? 12

What sells? 16

What are the long term Promises of 3D Printing? 18

Getting started - What equipment is needed? 20

Choosing the right printer for your budget 22

Visualize and draw your project 25

Supporting Plans 27

Become a 3D Designer yourself 29

Find 3D Designer Freelancers 31

Marketing your 3D Print 33

Starting a website 38

Introduction: Benefits of 3D manufacturing

What exactly is 3D Printing and what is the hype about it? 3D Printing is described as the fully automated manufacturing process of fabricating three dimensional solids from a digital image or model. It is also known as desktop fabrication or additive manufacturing. It is a process of prototyping whereby a real object is created from a 3D design. 3D printing has excited a lot of interest in the public domain due to the latest developments and the buzz created by social media. It promises a revolution in the foreseeable future in the fields of manufacturing as it provides a wide variety of manufactured products, including accessories, artifacts, jewelry, artificial limbs, prototypes for vehicle manufacture and an increasingly large bouquet of products.

The 3D printers, which are in essence the fabricators of the products, can create items very quickly from just a design or digital file to the actual object. This means the traditional manufacturing speed per unit is greatly reduced with 3D Printing and hence more rapid production of units per unit time is possible.

3D printers, which are in essence the fabricators of the products, can do so more quickly from just a design or digital file to the actual object. This means the manufacturing speed per unit is greatly reduced with 3D Printing than with traditional production methods.

Reduction in waste of materials

Traditional manufacturing is slow and wasteful. However with 3D printing, only products that are sold need to fabricated, therefore there is no wastage, no stocking up in a warehouse and so a reduced cost of production.

Creation of jobs

Engineers are needed to design and build 3D printers. A demand for technicians to maintain, supervise use, and fix 3D printers will be created and hence opportunities for jobs will increase.

Innovation in product development

Some of the innovative products that 3D Printing seeks to design in the near future include human organs such as the heart, lungs, or liver that will have next to zero chance of donor rejection, as the organs will be fabricated using the patients' unique DNA.

Hurdles in 3D Printing

Prohibitive costs; currently, 3D printers are limited with the size of the products that they can create. Most of the 3D printers capable of designing life size prints are still owned by large organizations. The smaller DIY 3D printers are continually being improved even as new manufacturing methods are developed though ultimately, large items, such as houses and buildings, could be created using 3D printers.

Limited materials

Currently, 3D printers only fabricate products out of a limited range of materials such as polymer, certain metals and ceramics. The use of a mixed materials and technology, e.g. circuit boards, is still under development.

Copyright of the technology

With 3D printing becoming more common, the fabrication of copyrighted material to create counterfeit items may become more common and nearly impossible to regulate.

Industrial waste problem

Generation of waste with the increase of 3D Printers even to the home: One of the dangers of 3D printers is that they will be used to create more environmental pollutants which are bad for the environment. Fortunately, there are new methods of automatically recycling objects made by 3D printers that hold promise of better recycling in the future.

3D printing technology is promoting mass customization of products and shunning traditional and obsolete manufacturing technology. With the implementation of 3D printing, as with all new technologies, there is bound to be a loss of manual manufacturing jobs that involve manual labor, which may in turn lead to loss of jobs. This may have a large impact on the economies of third world republics, specifically in Africa that are primarily estimated to be dependent on a large number of low-skill jobs.

3D printing brings with it advantages and marks a major achievement for mankind in bringing about the next phase of technological industrialization. It heralds the dawn of a new era in which manufactured products will be significantly cheaper and built quicker than ever before; however, the disadvantages of 3D printing need to be considered to be well studied, being in its early years, to be better understood and mitigated against.

How does 3D printing work?

There are several ways to implement 3D printing, one of them being direct 3D printing. Mostly, Direct 3D printing makes use of inkjet technology, not unlike the 2-D printing developed in the 1960s. It bears a similarity to the 2-D inkjet printer in that it is observed that in a 3D printer (though not necessarily), ink is dispensed via nozzles moving forward and backward bestowing a fluid. In 3D printing, however, the nozzles or the printing surface move in an up-down way to allow multiple layers of material to cover the same surface. Moreover, these printers use thick waxes and plastic polymers, which solidify to form a new cross-section of the solid 3D object. The end product of Direct 3D is a model or product with height, width and perceptible depth, unlike the 2-D print out. This is all sourced from a digital file in the computer that represents the model as designed by the CAD software.

Binder 3D printing uses inkjet nozzles to apply a fluid form a new layer, not unlike direct 3D printing. Unlike direct printing, binder printing also utilizes two distinct things that bind to form a printed layer: a fine dry powder plus a fluid adhesive or glue which is referred to as the binder. Binder 3D printers usually make two passes as they print to build each distinct layer. On the first pass, a thin coating of the powder material is delivered, and the second pass uses the nozzles to add the binder. Then the construction platform is slightly lowered to accommodate a new layer of powder, and the entire process is repeated to the completion of the model. MIT's 3DP process uses this binder approach.

Binder 3D printing has a few advantages over direct 3D printing. Expediency, as less of the material is applied through the nozzles than is the case with direct 3D printing, so the completion time of a process from start to finish is decreased. Secondly, you can incorporate a wider variety of source materials in the process, including metals, ceramics and pigments. So it demonstrates that you are able to create a wider range of prints and produce more composite products.

Another method of 3D printing is Fused Deposition Modeling (FDM). Fused Deposition Modeling (FDM) is an additive manufacturing approach that bears a similarity to direct 3D printing. FDM can create objects with features as small as a fraction of millimeter. This is the most used 3D printing method according to livescience.com. It is also becoming the fastest and most cost effective 3D printing method. It was invented in the 1980s. The co-founder and chairperson of Stratasys, Ltd was Scott Crump, a leading manufacturer of 3D printers. FDM 3DP is now trademarked under Stratasys, Inc. and works by inserting fluid plastic in closely packed lines using very tiny nozzles.

The most widely recognized printing material for FDM is acrylonitrile butadiene styrene or ABS. This is a typical thermoplastic that is utilized to make a great deal of basic buyer items, e.g. LEGO blocks. With the utilization of ABS, some FDM machines likewise print in different thermoplastics, such as polycarbonate (PC) or polyetherimide (PEI). Bolster materials are typically water soluble wax or weak thermoplastics, such as polyphenylsulfone (PPSF).

Photo polymerization and Sintering

Photo polymerization is a 3D printing technology whereby drops of a liquid plastic are exposed to a laser beam of ultraviolet light. The high energy beam of light converts the liquid into a solid, hence the term photo polymerization. Polymer is one of the multifaceted composition of plastic constituents with desirable characteristics to the 3D printing manufacturer.

SLA uses photo-polymerization; a photopolymer directs a laser across a vat of liquid plastic. In inkjet 3D printing, the SLA experiences the reiteration this procedure layer by layer until the print is done.

Sintering is a 3D printing/added substance fabricating innovation that includes the liquefying and melding of particles to print every progressive cross-area of an item. Specific laser sintering (SLS) is one manifestation of sintering utilized as a part of 3D printing and in this strategy a laser is utilized to soften a fire resistant named plastic powder, which changes its state to a strong structure to show printing. This looks like the component behind 2-D printers: They soften the toner with the goal that it will stick to the paper and make the picture.

Sintering can be utilized for building metal articles on the grounds that metal assembling regularly obliges some sort of reshaping and liquefying. The most commonplace case of using metal as a substitution of sintering material is an item called LaserForm A6 metal from 3D Systems. The articles made by the LaserForm A6 have a few focal points over metal items made by different means. For example, kick the bucket throwing the most unmistakable being the abnormal state of accuracy that SLS can attain to.

How 3D Printing Works
Despite the approach a 3D printer utilizes, the general printing procedure is pretty much the same. In their book "Additive Manufacturing Technologies: Rapid Prototyping to Direct Digital Manufacturing", Iaan Gibbson, Brent and David W. Rosen record the basic steps followed.

Stucker records the accompanying eight stages in the bland AM methodology:

Step 1: CAD mm Produce the 3D model utilizing PC helped outline (CAD) programming. The product gives signals as to the auxiliary uprightness expected in the completed item, as well, utilizing exploratory information about materials used to make virtual recreations of how the article will act under given conditions.

Step 2: Conversion to STL. Conversion of the CAD attracting to standard tessellation dialect is a document configuration created for 3D Systems in 1987 for utilization by its stereolithography device known as (SLA) machines. All the 3D printers presumably utilize STL records alongside some restrictive record sorts, for example, ZPR by Z Corporation and ObjDF by Objet Geometries.

Step 3: Transfer to AM Machine and STL File Manipulation. The client duplicates the CAD produced STL document to the PC that controls the 3D printer. The client then assigns size and introduction for printing; pretty much the way you would set up an ordinary desktop paper printer.

Step 4: Machine Setup. Each machine has its own assets for how to get ready for another print work. These are as per the following refilling of folios, polymers, and different utilities that the printer will utilize.

Step 5: Build. Letting the machine execute the occupation; the assemble methodology is basically programmed. Every layer is more often than not around 0.1 mm thick, however it can be somewhat or much more slender or thicker relying upon the model's size, the 3D printer and the materials utilized, this procedure can run for a considerable length of time or even days to full fulfillment. You ought to wiretap the machine intermittently to verify there are no lapses.

Step 6: Removal. Taking out the printed item (or various protests now and again) from the 3D printer. Security safeguards, for example, wearing gloves are important to shield yourself from hot surfaces or lethal chemicals and to dodge harm by and large.

Step 7: Post-transforming. Many 3D printers oblige some machining/post-handling for the printed item. To uproot water-solvent backings incorporate getting over any exorbitant powder or washing the printed question in water is utilized. This step may have powerless new printing since a few materials need time to safeguard, so care ought to be taken may be important to guarantee that it doesn't break or disintegrate.

Step 8: Application ought to utilize recently distributed protest or items.

These steps are the basic procedures in creating a print from its ideological stage to the realization of a 3Dimensional object for a customer or client.

Is 3D printing profitable?

3D printing is quickly going mainstream and as a result, many innovations are coming up in this technology to enable it to meet the needs of customers/consumers.

3D printing has grown at a very high pace, with recent statistics showing 3D printing competing closely with traditional manufacturing. With smart choices, you can benefit from the 3D printing revolution which is slowly changing the traditional business model with many advantages being realized, such as the elimination of traditional factories, recruitment in 3D printing and manufacturing, design freelancing, and consultation as well as product sales. This makes venturing into 3D printing opportune at this time so that you can place yourself in line to reap the numerous benefits of the industry.

Though creating sellable ideas may be a challenge at first, there are online forums where one can get ready ideas for discussing modifications and tooling. There you can consult, learn and find out more about 3D printing from already established experts for free, then move on to try out their suitability in online groups.

Currently, you may not need to purchase equipment to print as there are 3D Printer owners who, through the internet, will allow you to send them your design which they will then fabricate or print and charge you a fee and ship the finished print. This can be especially affordable as a start-up, and this will buoy your confidence as each idea is turned into a finished product. Meanwhile, ideas are as unique as every human has their own mind, thoughts and ideas. You can create your own niche and begin a profitable venture through simple designs. If you wish to move to designing the ideas you have and perhaps have no prior knowledge of designing, again you can find many freelance designers who will, for an agreed fee, convert your idea into a model without breaking a sweat. This will save you time which you can use to come up with new ideas and the perhaps complex designing software learning time and cost of the actual software.

Below is a brief guideline on how you can gain a foothold in all the potentially profitable areas of 3D Printing:

You can begin by supplying 3D Printers & Providing Printing Services. Selling 3D printers is a sure method you can use to earn a neat profit out of the 3D printing revolution. 3D printing technology is now quite popular. It has many people pondering buying their own 3D printers, which means that basically someone has to come in and fill this void as there is a prepared marketplace for 3D printers. That is making corporations prefer 3D Structures, and Stratasys progressively famous and lucrative. The cost of 3D printers has gradually decreased, and 3D printers have become affordable to patrons and 3D fanatics alike. The 3D printer is now, more than ever, becoming accessible to consumers. This is a solid reason why one should venture into sales of 3D printers.

Provision of Printing Services to clients can be greatly profitable. Many 3D printers are reasonable, with most printers priced at about $2500 or more. You can start your own printing business simply by buying your own printer.

Selling Customized Goods

There are a wide variety of 3D printing consumer products and belongings, for example; figurines, key-holders, wristbands, watch cover cases and smartphone carrier bags amongst others. All it takes is creative imagination to plan, print and vend your personal, exclusive modified products. 3D printing can propel you to be a "work from home" businessperson by just designing an original and unique product which can then be marketed.

Selling of 3D Printing Supplies and Software

Due to the many people today who want to own and use 3D printers, you can earn income by vending 3D printer supplies. Supplies for 3D printers can include 3D printer cartridges to the Graphene and pigments for 3D printing materials.

Providing 3D Designs Online

If you are talented in professional design, you can sell 3D designs and models online which will give you an avenue to profit from the 3D printing revolution. These skills are also provided online and you can get proficient with the simple-to-use design software and begin to provide services of designing models for customers who may have ideas but lack the technical know-how to implement them. What unique designs you may have created can really boost your profile as a designer if you post them on your social media page. You can do online promotion by sharing site links with friends and colleagues, telling them that you provide some cool 3D designs through your website. As long as your models are stunning and captivating, it is not difficult to find people who will be willing to buy your work. With the large number of 3D-devotees who lack the sufficient resources to design for themselves, buying 3D designs is easy and simple.

Printer Repair Services

Like any other machine, support services are crucial for 3D printers to do required repair or maintenance services. If you can acquire the necessary skills to tackle the maintenance and repairs on 3D printers, you may be able to enter the profitable market of 3D Printer services. Your role will essentially be fixing 3D printer problems. With just the core skills, you would be able to establish your business and advertise online to reach out to your customers. There are online forums on advice and FAQs on 3D Printing which you can take advantage of to discover invaluable tips on 3D printers and other things that might lead to a better consumer experience. You will begin to attract traffic to your website and soon people will be flocking to you for your services.

Investing in 3D Companies

Last but not in any way the least, you can benefit from the 3D printing revolution by investment in leading companies in 3D. You can buy stock from well-established 3D companies such as Stratasys, Organovo or 3D Systems.

What sells?

To get into the 3D business you will need to know what the top selling products on 3D printing are. Below is a list of the fastest moving products from some of the leading 3D printing retail outlets.

The Moto 360 Bumper case. This is a brilliant design that exudes simplicity made for a very popular smart watch: the Moto 360. The bumper casing is used for protecting the watch from mechanical blows. It is raised 1.5mm over the front glass, to protect the watch on all sides from normal wear and tear. It fits like a glove on the face of the watch and doesn't interfere with the microphone or crown button. It is designed to accommodate the charging dock. It is the number one selling item because it comes in a variety of colors.

The Pouting Keanu Reeves

This is a miniaturized doll of Keanu Reeves that depicts the actor in a slouched or slumped position. It was designed by neural firings, and is quite popular. It is fabricated in full color sandstone. It can be bought in two distinct sizes starting at $25.00.

The H3 gimbal GoPro bracket

This is a modified version of the screw-less bracket for the GoPro Zenmuse. It also provides for screw-less support, in that it allows for easy clip on and off. It retails at $11.79.

The FitBit Flex is a wristband that keeps the details of your jogging or track activities such as the number of steps you take, the distance you pass in how much time, the number of calories burned, and the quality of sleep, and it can noiselessly rouse its wearer to full wakefulness. It retails at $8.99.

The Gridded Stereographic Projector

This is a spherical object that is on sale on Shapeways website. When light is shone from directly above the sphere, it projects a gridded layout over the surface. The 3D print with spherical design can be bought in 7 different color options for only $17.00.

NightscoutDexcom-Moto G Case

This 3D printed Moto G phone case that allows for the attachment of a Dexcom G4 glucose is actually the monitor receiver, making the device easy to use and less burdensome. It retails at $60.

These are just a sample of the best-selling products that are listed on the website Shapeways.com, that also includes detailed descriptions of the products. The products are simple but unique and well designed to catch the eye of the consumer and go to show the effectiveness of selling simplicity. But no one can deny the fact that the more practical and uniquely aesthetic the application of 3D Printing, the easier the sell. This means that anyone with a feel for art can begin to explore the market of 3D printing with a high chance of success based on their own artistry and originality.

What are the long term Promises of 3D Printing?

The promises of 3D Printing to the future of the manufacturing industry, construction, healthcare, education and even commerce and industry are currently being studied and researched as the technology continues to be embraced in all the aforementioned sectors. The current development in 3D Printing is only the tip of the iceberg as pertains to the potential that 3D Printing technology has to offer. I will endeavor to list a few that are currently being pursued by the main player in all sectors of the economy

Small Batch Manufacturing

This is one of the futuristic goals that has already been achieved to some extent. With the growth in home-based 3D Printing, small batch manufacturing is being realized right now, though on a diminished scale. Once the prices of medium to high tech 3D Printers come down as projected, and the materials and printer are made readily available, then this will be widely achieved. The upside of this is that 3D printing undermines different laws of balance, and the reduction of waste production in the manufacturing sector. This has the two pronged advantage of affordability and hence marketability of the end product, and also environmental friendliness due to limited waste production.

Bioprinting

3D Printing promises to create products of future healthcare including kidneys, living skin grafts, advanced prosthetics, and dental implants. There is a lot of expectation riding on the potential of 3D printing to realize the production of these organs.

There are serious challenges facing researchers as they attempt to enter into the bio-printing field. Research is on-going to achieve this in the near future. It would really be a great leap forward for humanity if living organs could be printed that could replicate the durability and resilience of natural human organs.

DIY/Open Source Revolution

In the market currently, the number of 3D printers is very low. Their capability is also relatively limited in terms of the precision and the number of materials that can be printed. The potential, nonetheless, exists for the continual discovery of more accurate techniques and a wider variety of printable materials. This is the promise that is a function of time and resources. The upside is that technological innovations are rapidly being advanced daily. Compare this to the hall-size first vacuum tube transistor computers, only 30 - 35 odd years ago which have become today's smartphones with high processing capability. This is one promise that is coming closer to reality with each passing day.

Printable Food

This is a promise that has a lot of challenges in realization, but just like bio-printing, may have tremendous returns if achieved. Currently, the technology is restricted to confectionery and pastry. If healthy foods, not unlike protein bars and vitamin pills, could be printed in 3D printers, the cost of nutrition and healthcare could be reduced significantly. The potential gains are not hard to see and indeed this remains a very loaded promise.

Getting started - What equipment is needed?

The requirement for 3D printing is very simple and minimal and is the key propagation catalyst for home-based 3D Printing that is rapidly gaining popularity as a start-up venture.

To begin with, you will require a desktop computer that is loaded with the Computer Aided Design (CAD) computer application software that will enable you to create/develop. Anything manufactured within the last 3 years will be capable of handling the CAD program requirements and no additional processing capacity will be required unless guided by the installation requirements.

Secondly, you will require an actual printer where you have in mind product production as opposed to only 3D Print service. The equipment may come assembled or in component form. If it is not assembled, you will need to assemble it as per the instruction manual in order to prepare for operation. Sometimes the printer comes with the print material as one package. If the material is not included, you may just have to source it separately. It is best to find out as you are purchasing the 3D Printer what is inclusive in the package and what is not.

Thirdly, you will a connection to the internet, either through a cable modem, wireless modem or just Wi-Fi that links your computer to the internet and also links your computer to the 3D Printer. The connection to the internet is vital as you may need to do 3D Printing as a service and customers will need to reach you online.

Last but not least is the normal electronics guard equipment and Unlimited Power Supply (UPS). This comes in handy when black outs and brown outs are experienced. The 3D Print process requires that the power supply is not disrupted. That summarizes the ease of set-up of the 3D printing equipment and setting up your business.

Choosing the right printer for your budget

For the start-up business in 3D Printing it is critical to get the right machine for your venture of interest. There are a several brands that are currently available and the choice is strictly to be determined by your budget. Though 3D Printers and the print material are expensive, the cost has dropped a great deal in the last few years. Below is a list that bears a variety of printers and their current prices.

The Buccaneer Cloud 3D Printer, billed as a success for Kickstarter, the first 3D printer was sold for $350. The manufacturer's purpose was to make the 3D printer the most user-friendly, cost-effective and innovative 3D printer on the market.

The Buccaneer is built to be set up easily and ready to print just from the box and is controlled by using mouse buttons to manipulate and design basic objects in any way you desire. Upon completion, you can share your image with your friends on-line or send it directly to The Buccaneer 3D Printer and wait as it is made!

The RepRapare 3D printers are relatively low-cost and available in versions which vary in price from about $700-$1100. It is a unique general-purpose 3D printer with self-replicating capabilities. Many of its parts are made from plastic and RepRap prints those parts. RepRap can fabricate its own copy by making the plastic parts that it is made of. The printer allows you to fabricate many useful items over and above its own plastic components.

The Cube 3D Printer is a consumer-friendly printer that is available in a wide variety of colors. It is user-friendly which means it is designed to be particularly easy to set up from the box. You can connect your desktop computer to the Cube via Wi-Fi and at $1300 is more costly that the RepRap but has superior output quality.

The Up! 3D printer is not unlike the Cube since its manufacturer set out to endear 3D printing to the masses with user-friendly features. It costs $1000. A higher version called the Up Plus and Up Plus 2 can be bought for $1600 and $1800 respectively.

The Makerbot Replicator is a 3D printer which can print high-resolution, smooth objects quickly in a variety colors, and can produce interconnected components as well as moving objects. A smaller one costs $2200, the largest $6499.

It is driven by the fresh, comprehensible MakerBot Replicator 3D Printing Platform and is Application and cloud permitted. The 3D printer is also Wi-Fi, USB, and Ethernet compatible, which guarantees a unified manufacture workflow. The designer can even access the Makerbot Replicator with the above mentioned network technology, to remotely connect to, view, and control it.

The CubeX is manufactured by the manufacturers of the Cube. Its functionality is to print giant models with peak resolution and even make skilled ranking parts. It produces various prints of multiple items over the single surface. Different colors and different quality plastics made at the same time sold for $2500.

The Printrbot is a 3D printer with high capability and is usually for the printers who indulge in 3D printing as a hobby. It is also suitable for a start-up 3D Printing business and can be bought fully assembled or in component kits which you can then put together yourself. The fully assembled Printrbot starts around $399 to $699; it costs $259 to $299 for the unassembled version.

The above are a few prominently marketed 3D printers but the list is by no means exhaustive. There are other 3D printer models out there. Depending on your budget for your 3D Printing venture, you will be able to select from the available 3D Printers to best suit your business.

Visualize and draw your project

A project's success or failure is determined by the planning. You will have to create a solid plan for your project before undertaking any activity in the project. For a start-up, this may be difficult especially because the 3D Printer entrepreneur is raring to dive headlong into the actual business. It is crucial to get it right by having a working and reviewable plan that will monitor the use of vital resources and time. Below is a look at the simple, practical approach and crucial exercise of project planning.

Project Goals

The first step of any project is setting goals. These goals are guided by what it is that you need to meet as a stakeholder to declare the project a success. To begin with, identify the stakeholders of the project. It is difficult to recognize them, especially those who the project will impact indirectly. These are usually the sponsor(s), customers who will receive the deliverable of 3D Printing, and users of the project outputs. Last but not least is the team you are engaging, if any, to accomplish the objectives. In the event that you are the sole sponsor and therefore stakeholder of the project, you can proceed to enumerate on your needs. If you are not convinced that the project will bear you the desired benefits, scrap it.

The next step, once you have recorded a comprehensive list of needs, is to prioritize them. From the prioritized list, choose and classify the goals which can be easily gauged. This is where the SMART principle comes in. This way you can easily tell when a goal has been achieved. Once you finalize the goals, they should be the active part of your project plan. This is the toughest part of the planning process completed. Up next are the project deliverables.

Establish the project deliverables

With the goals you made above, you have to generate a list of things the project needs to provide in order for the goals to be met. Include the stage of delivery and the mode for each item on the list. Now incorporate these deliverables into the plan with the projected delivery date. More accurate delivery dates can be fine-tuned during the scheduling phase, which follows consequently.

Draft the Project Schedule

A list of tasks that need to be carried out for each deliverable identified above has to be created. This needs to be done hand-in-hand with establishing time needed complete the task. Once established, you can work out the effort required for each deliverable, and an accurate delivery date. Review the segment on deliverables to acquire better delivery rates of your own choice.

You will need to use a software package for scheduling of events at this stage of planning. Microsoft Project is a recommended tool that comes in handy. You can also choose to use one of the many free templates available on the internet. All you have to do now is to plug in the details of all your deliverables, tasks you are going to perform, timespan of work and the approaches required for every task to be completed.

You may experience challenges with keeping to the time schedule. This may be occasioned by delayed delivery due to skills gaps and equipment acquisition & shipment. The project schedule should guide you in adjusting the delivery date due to unforeseen factors.

Supporting Plans

The planning process facilitated by this section deals with plans you should create as part of the overall plan. You can easily add them to your project.

Human Resource Plan

Once again, you are it! You are entirely responsible for the project unless you hire someone to assist you in the business. For records' sake, you can outline your roles and responsibilities as the project progresses. Also estimate the resource start dates, estimated duration and the method you will use for obtaining them. Create a single sheet containing this information.

Risk Management Plan. Have a risk management plan to identify as many risks to your project as possible, and be prepared if something bad happens. There are many risks, below are a few:

- Hopeful about time and cost estimation
- Inadequate capital
- Unexpected price fluctuations especially of materials and equipment for 3D Printing
- Stakeholder requests deviate from the initially described requirements when the project has been started
- New requirements cropping up after the project has started, e.g. regulations and licensing

By creating a risk log, these risks can be tracked and managed by pre-planned action that can be taken when it occurs and steps to prevent a repeat occurrence. This risk log should be reviewed periodically during the project and all risks identified must be addressed when they occur.

On following the above steps, you will be well on your way to achieving the goals of your 3D printing project and you will realize the end in the absence of a calamity. It is advisable to constantly update your plan as the project progresses and measure your achievements against the targeted results.

Become a 3D Designer yourself

It is becoming more affordable to join the increasing number of 3D entrepreneurs as manufacturers continue to produce new, low-cost and easy to use 3D Printers. The creators of RepRap 3D printers had in mind the affordability aspect and ease of set up of their 3D printers in order to popularize them. Though 3D printers are easy to access and use through a connection to a desktop computer, a user may need to be able to design and create sellable models. They may also need to join online groups which are gaining in popularity as a forum for the exchange of ideas and platforms for FAQs on the various challenges the individual start up may encounter.

These 3D Printer manufacturer forums feature the manufacturers of actual 3D printers who use them to answer the questions of their customers and prospective clients. They also put their heads together as manufacturers to tackle common barriers such as copyright laws and legislation pertaining to the use of the materials for manufacture and their environmental impact.

This may be a good place to start as a DIY designer; getting to know the industry leaders and the reviews on each manufacturer and their market share. This will help you get a picture of the kind of support available for the particular 3D printer that you intend to use. Once you have established the product you want to use, you will want to visit the relevant sites.

What do you need to know/skills?

You will need knowledge of basic computing to begin with. This is so you can use the DIY 3D printing software in order to practice and develop designs and models to print. You may visit

3D forums that are based on freelance designing for 3D printing which may help you gain the latest and most relevant skills that are required in the market.

What resources are required?

First you will need to have access to a desktop computer or any portable device that can allow you to manipulate a design into a 3D printable model.

You may also require a scanner that can take a 360 view of an object that you wish to replicate.

You will require Computer Aided Design (CAD) software. There are some free software versions that are very easy to use which are also available on Google and there are professional/commercial version of CAD software that you may require training for in order to gain proficiency.

Access to the internet

The internet will enable you to access online forums through which you will receive job orders and web-based assistance. You will also use the internet as your chief marketing tool once you get adept at designing for 3D printing.

Where can you get training?

Training can be accessed at technology colleges formally, but informally, the online forums for 3D print designers will point you to tutorials and trial programs for gaining proficiency. There are also freely downloadable tutorials on Google and the internet at large.

Find 3D Designer Freelancers

There are many websites on the internet that list 3D Printing designers for hire, but they basically all give the same parameters for listing these designers. In essence, there are three vital details which determine how 3D Designer freelancers will approach the job. These are the following.

Clarity on what is required of the designer

It is vital to be very specific about the intended end product. The more clear you are in putting across what you need, the more likely it is that the designer will understand from the word go. Supporting your idea with sketches, photos, magazine clippings and even screenshots of elements you like are all really helpful in communicating what you desire as the final outcome. Things like style, final texture, dimensions and materials to be used will also go a long way in guaranteeing that you get the right 3D print.

Communication

Designers are guided in their creative nature that brings life to ideas through effective. For this to take place , there must be open and frequent communication to ensure that that designer has a clear understanding of your requirements, and you have knowledge of their timetable. They should be asking you just as many questions are as you are listing specifics. You must at all times be polite and honest in order to get the best service. Use positive language highlighting what you want and describe efficiently what you like and what result you require. It is also highly recommended to use descriptive language as much as possible and to avoid minimalist statements. On communication, it is also vital to communicate the terms of engagement so that no party springs surprises in the middle of the job. It is best to have a formal agreement on record to which both parties must consent for the 3D Design to get underway. This agreement should also reflect on the timeline for delivery of the product and the cost.

Cost

The cost of the job will be dependent on certain factors, which are listed below.

Time and labor

Large and highly detailed jobs take longer and require more labor, hence a higher cost implication. Will the deliverable be a finished product or a 3D file? If you just need a design image to print for yourself, it will be much cheaper than if you need the design model printed due to cost of materials

Design uniqueness

For a one of a kind item, you will pay more as opposed to a common item due to the creative energy involved in bringing your idea to life.

Similarity of item. It is best to look at the cost of similar items before you engage the designer to get a feel of the price range.

These three Cs as outlined by Shapeways Freelance 3D designers' blog will guide you when you need to hire a 3D Designer.

Marketing your 3D Print

In order to successfully reach as wide an audience as possible, you need to embrace the power of the internet. It gives you innumerable advantages in free and premium advertising that is simply not achievable through any other medium. The internet is also interactive, instant in communication and offers a plethora of professional services that could help your 3D Print business to thrive. It is therefore very important to exploit the internet to your greatest advantage to reap rich returns from its vast potential. The following are guidelines of how you can harness this powerful medium.

What is your niche? You need to identify your customers, identify what they want, what their mindsets are and more importantly, their buying trends. You may need to research further your intended customers to learn how to rope them in and get them buying your products or services.

What is the marketing plan? A marketing plan helps you to come up with and execute sound strategies for increasing your visibility and accessibility to your intended customers. It also helps you to monitor the success of the adopted strategy and finally to evaluate your progress. It should highlight all measurable objectives and establish milestones that will give you a pointer of where you are heading as an entrepreneur.

Have you ventured into social media? Social media is a an increasingly powerful ally for internet advertised businesses that can be harnessed for marketing first to a circle of colleagues and confidants and then later to their circle of influence - friends of friends with a vast potential for raking in profits due to the large volume of sales achievable on it. You can create a page advertising your services and have customers give testimonials which will in themselves be a powerful selling point in addition to graphics on your page cataloguing your products. Once you know your audience, you should replicate your profile on Facebook, Twitter, LinkedIn and the likes to reach as wide an audience as possible where your customers are to be found.

Do you have a professional website that is SEO & SEM enabled? A professional website forms the entry point into the internet market and is the face of your 3D printing business before clients get to know you. It needs to have a simple to navigate interface that can allow for feedback, and FAQs to allow your customers to interact with you and get familiar with your products. It is invaluable to your marketing strategy, and enabling the Search Engine Optimization features on it will ensure you get more hits by visitors any time they enquire on a search engine about your product. You will need to have the website designed by a professional for SEM and SEO features to be present.

Are you acquainted with email marketing? This is a highly effective and largely inexpensive tool of marketing that will aid in achieving your business marketing strategy. It is much easier to send promotional mails to your network of established customers than it is to send snail mail, and is definitely more cost effective.

Other critical factors that you may need to consider are your competitors; knowing what strategies are working for them and the market analysis tool that they are deploying in order to keep ahead of the pack.

In summary, the key to selling your 3D printing business lies in embracing the time-tested marketing strategies that harness the full potential of the internet.

Social Media

Digital marketing has experienced a meteoric rise in recent years, and this has informed the decision of many businesses to take advantage of social media for promotion of their products and services to potential and existing customers. The most popular social media are Facebook, Twitter, YouTube and LinkedIn, which are heavily used by many businesses for advertising, interaction, surveys, feedback and general infographics. This is by far and large the most popular marketing tool as compared to the hitherto advertising leaders, mass media. Here are the basic advantages that one can get from advertising their 3D printing on social media.

Social media has a large audience

Adverts on websites, newspapers, TV or magazines usually have limited access and may in many cases be subscription bound. In contrast, social media sites give 3D Printing business owners access to a large audience who can choose to "like" or "follow" your business and its updates. These features, especially on Facebook and Twitter, encourage advertising through peers where people with links to an individual can see his likes an follows and can also follow suit. Facebook's estimated users are 750 million. LinkedIn users number around 120 million, and YouTube is around 3 billion, YouTube videos are viewed daily. This can present a boundless client space for any 3D printing entrepreneur and give him/her a good chance to reach clients on a global scale.

Social media draws attention in real-time

Adverts in periodicals, dailies, magazines and even sometimes TV take a while to reach the intended audience. This is not the case with social media. You are able to do social promotion of your events as soon you finalize plans, and social media will advertise immediately. Social media gives 3D printing business owners the ability to expand their reach with the help of blog posts, introducing tips and ideas, and providing coupons and competitions to involve people all around the globe.

Increase traffic to your website

Sometimes due to SEO logistics, the traffic being directed to your website may be low and social media marketing can be the vehicle to boost the number of people who visit your site, how long they stay on it, and encourage them to keep coming back. You can use social media promotion to increase your web traffic by posting links to products and services you offer on your website. You are also able to stay in connect with your fans and followers of your blog.

Advertising that is interactive

You may want to periodically carry out a survey to find out which particular product your customers favor, or to ask customers to subscribe to your new email catalogue of your latest 3D prints. This is all made possible through social media. It allows you to interact through message, chat and even forums. You may even ask a question and watch how quickly your fans and followers reach out to give you their responses. Customers can even leave messages on your page if they have a customer service concern. Social media marketing helps to provide online customer and business interaction. Social media networks also give you a chance to engage with other like-minded businessmen and industry leaders in online groups formed through social media. The information you receive can help you improve the way you manage, operate, market or finance your 3D printing business.

The medium is affordable

Whereas traditional advertising methods like television and radio promotions, banners, handouts, poster placements and participating in promotional events can require a large proportion of the running cost of the 3D printing business, it's free for businesses to sign up to popular social media networks. Social media is the way to make loyal customers, and an affordable way to promote your products and services.

Customers feel like they have a way to access you if they have questions, and feel like you genuinely care about their opinions. In fact, they become your true customer. Social media marketing helps you to engage customers and succeed in building long term relationships. When you post any update, video, promotion or sale they keenly respond and show their interest. Social media is therefore a vital component of advertising for anyone who would like to venture into 3D Printing.

Starting a website

Is creating your own website the way to go about marketing your 3D Printing products and services? Well for starters, a website is one way of advertising for your 3D Printing products and services on the most visited library in the world - the World Wide Web! It has three particularly vital advantages for your business.

It is a cost effective method of doing business

Say you have a limited marketing budget, advertising on your own site is a low-cost means of marketing. You are free to determine the materials on your site, the colors, the images and the catch phrases that appear on your website, allowing you the freedom of creativity and exclusive ownership. You only have to pay the hosting server provider for your site and the maintenance team if you are not maintaining it yourself. These guys are normally paid periodically and the costs are manageable.

Potential to increase your visibility to new customers

Web advertising greatly increases awareness of your company and reaches a whole new set of potential customers. Folks who may not even know of your retail location can become ardent online shoppers who enjoy shopping on your website. An effective and healthy way of ensuring a loyal and satisfied following can be, for example, offering these folks discount cards. This will go a long way in encouraging your in-store customers to visit your website, which will invariably boost sales. This is what you want as a start-up or even as an established business - to safeguard the all-important bottom line.

Websites provide a Personal Touch to your 3D Printing business

Normally the internet feels like a very impersonal place, but having a website can be a way for customers to get to know a little about you and to give you feedback. Once you establish feedback mechanisms on your site like a contact space and feedback column, you care able to keep in touch and, more importantly, in tune with the customers' needs. Some buyers surf the Web to research a product they are considering buying. Your website can provide details about what you are about and your product catalogue. It's often nice to find testimonials from satisfied customers and to get to know the face behind the website and the business through the About Us page.

Though a website has these three undeniably powerful advantages, it is indeed critical to have all the right ingredients on the website in order for it to be effective and impactful. It should have the qualities below.

The website should be focused on usability rather than beauty. Beauty does not always translate to sales. You want a website that will tell visitors or customers your core business in 3D Printing and make it easy to make enquiries and purchases with full information on products/services that you are offering.

It should be easy to navigate from your sitemap, through links and images. It is good practice to use technologies that do not bring about accessibility problems across the various browser platforms that are available. Flash media and Java scripts can cause these problems; use HTML as it is easily displayed by all browsers.

Sometimes there can be too much clutter on a website due to an over-zealous developer. It is good practice to keep to a bare minimum use of audio and video. Always ensure that you use compressed & easy to load files as a rule. Larger files can slow site loading speed, and intolerant visitors often leave slow sites. Ensure you get a skilled graphic designer to produce an attention-grabbing header for the site. It is usually the first "impression" that your visitors will see, and therefore must be impactful.

It will also help immensely to use powerful headlines for your site

When the powerful headline is accompanied by complementary images, they have the desired effect of catching and holding the attention of the visitor long enough to register interest and act on the information by either enquiring further or making a decision to purchase.

Advertising on a website is a good way of putting the message out there about your 3D business, and will boost your business within your locality and beyond. Though the points I have outlined boost your profile to your potential customer, the addition of Search Engine Optimization features will also boost visibility and draw even more customers your way with the rapidly increasing number of businesses in the area of 3D Printing.

One of the biggest promises of 3D printing is the ever expanding field for which it can be practically applied to a diverse population of consumers. From engineering models for rapid prototype development to dental models, prosthetic limbs for amputees to candy making, ornament making, and even toy making, more users and consumers are taking part in the 3D printing revolution as it is becoming more relevant to their lives. Now more than ever, there are sites through which a person can hire a designer, send them their idea and have a workable model designed for a fee. This means that even people who may never come into contact with a 3D printer are in a position to influence a design to their liking and make 3D Printing all the more popular.

According to analysts at Juniper Research, under its new report Consumer 3D Printing & Scanning: Sales of consumer 3D printers are set to climb to more than one million units by 2018, according to the paper, from just over an estimated 44,000 this year. That marks an exponential growth in demand and also product availability from the increased number of 3D Units.

Report author Nitin Bhas said that sensitizing the public and educating them on the idea of creation of everyday objects through 3D printing will be important to ensure it has success in the long run. He pointed out that novel, revolutionary apps and creative content would give impetus to the 3D revolution, helping it bring about custom-made products honed to personal taste that are not yet on the market.

The automotive industry largely requires prototyping as the base for modeling new versions or new models of automobiles to go with the times. They basically produce as many prototypes as they produce authorized designs. 3D Printing makes the production of prototypes easy, more cost effective and faster to produce. The motor vehicle industry also requires the manufacturing en-masse of vehicle components, parts and accessories, and all these are easily created via 3D Printing.

With the help of 3D printing, the manufacturers are able to consolidate many components into a single complex part, saving on time and materials since 3D Printing also prevents wastage of the raw materials used, unlike in the methods previously used. It also reduces the amount of waste from defective design being a more accurate manufacturing process guided by CAD software.

Moreover, 3D printing can also be used in production tooling and can yield more accurate results with the use of rapid prototyping form and acceptable testing.

In the aerospace industry, 3D printing is becoming indispensable

Now complex parts that were hitherto impossible to fabricate with traditional manufacturing are considerably easier to create, more so with the parts that have complex geometric designs. With 3D printing, the use of composite compounds, polymers and others, the material properties of the aerospace product may be varied in density, tensile strength and other material properties to produce the desired result.

In the pharmaceutical and healthcare industries, the use of 3D Printing has enhanced the creation of educational anatomic models of the human body which have aided understand the workings of the human body and the precise functions of each part. Through the use of digital imaging from CAT scans and ultra-sounds, CAD programs have been able to model organs that are under observation to aid doctors and surgeons in recommending the most appropriate surgery or surgical treatment and therapy.

Using 3D printing, Orthopedic implants and prosthetics have been and are being developed that are well suited to each individual patient's need. This has led to an increase in effectiveness of these artificial limbs by increasing their suitability and best fit.

The retail and commerce industry has not been left out of the market share created by the introduction of 3D printing in the manufacture of consumer products.

Now toy manufacturers can create custom toys

Jewelry can be directly printed through 3D printing without going through the rigorous tooling process which can affect the price of the jewelry.

Board games like chess and draughts can now be created to fit different audiences and their preferences in art or symbolism through the unique shapes achievable via 3D printing.

Another retail revolution is the whole new spectrum of home decorations, from artificial plants to miniaturized plant pots, 3 dimensional stand-alone artifacts, mugs, vases, etc., which 3D printing has made possible. The designs are as unique as one human being to another.

3D printing will also greatly affect the way we approach the manufacture of sports equipment. 3D will be truly phenomenal in the manufacture of items like sports shoes, sports balls, trophies, and more. This really showcases the ability of the printing equipment by creating complex geometry and shapes not possible with traditional manufacturing. The geometric shapes and artistic shapes are only limited to the imagination of the manufacturer and the limit of the 3D Printer's capacity. This was not the case previously, where the mold and material would be the sole determinants.

Manufacturing protective safety gear that is an exact fit

3D printing has also enabled the manufacturer to create a more accurately fitting glove, pair of goggles, snorkel, etc., where the previous technology was limited in streamlining and perfect fit. Now, ski boards, skateboards, ice skates and others can be made to precision that allows maneuverability and aerodynamics without compromising on protection of the wearer. All these are created from the availability of bio-mechanical data.

"3D printing has the potential to revolutionize the way we make almost anything."

~President Obama

Three dimensional printing is indeed the future of manufacturing and is the very next phase in the evolution of customized product development. It has achieved this in the few short years it has taken to popularize it, even though the initial cost of design implementation was relatively high. However, prices for 3D printers have continued to fall even as simpler and more cost effective manufacturing methods are being developed for the production of home version 3D printers. Social media is playing a vital role in creating technological awareness and exciting public interest in 3D printing, so the future looks bright for all who are thinking of venturing into this line of manufacturing. The entrepreneurial opportunities for DIY venture lovers expand daily and offer endless possibilities. This technology winning both appreciation and satisfactoriness will greatly influence the way we see, think about, approach and tackle current and future world problems. 3D is truly revolutionary.

www.ingramcontent.com/pod-product-compliance
Lightning Source LLC
Chambersburg PA
CBHW030517220526
45464CB00006B/2833